THE ROCK STAR
SCRUM MASTER'S MANUAL

Launching Kick-Ass Scrum Teams

MATTHEW KRAMER

Printed in the United States of America

Published by:
AAE Press
First Printing 2023
First Edition 2023

ISBN 979-8-9862060-2-8

Cover by Tanja Diamond

This book is dedicated to my sister Jenny Kramer who has always been a great friend and ally, and to Tanja Diamond who helped me turn this idea into a well written book that I'm proud to deliver into your hands.

TABLE OF CONTENTS

INTRODUCTION

If you have been working with the same team for years, or if you're about to start up with a brand-new team or even start a team from scratch, there are great things to learn in this book. I have launched scores of teams in my career and taken over lots of existing teams, including some of the most challenging organizations to be had. I won't promise that my methods will be the easiest approach, because they aren't. The easiest approach doesn't require any work, but you'll get poor results and pay for it for a long time. My methods need an upfront investment, but I can promise you that investing up front isn't as hard as you may think, and it saves you months or years' worth of the team growth and personal effort required for you to get there.

How To Read This Book

This book exists for two reasons:

Reason 1: This book is here to get you great results by giving you the information you really need to be successful on the ground, at the actual job.

Reason 2: It's here to make the standard Scrum certification course look as sad and inadequate as it really is, so that when you need training, you'll think of me and the follow-on course that has been created from this book (available on-line or in person).

Here is the situation that we're trying to avoid.

As a new Scrum Master, you're leading a high-profile team responsible for delivering new content for an important client who brings in ten million a month in revenue. Your team has been working hard and there are two hundred plus people on the zoom including your boss and the CEO so it's your time to shine.

You start showing off the brilliant new functionality that the team has been working on for the past three sprints when halfway through the demonstration the application crashes, and you notice a glaring spelling error in the product. When you finish, the director of sales mentions that he didn't see critical functionality promised to the customer. Your victory lap has been crushed in front of two hundred people, your boss is angry, and the Development Team that worked late nights to get this across the finish line is now demoralized.

I've seen this and worse at companies that I've worked for. Fortunately, this situation is avoidable and it's why I've written this book. This book is a practical guide for current and future Scrum Masters on succeeding in the real-world, day-to-day execution of the SM role. In short, this book will teach you how to be a Rock Star Scrum Master.

What this book is not

There are plenty of books and training material about the basic Agile theory and concepts. This book does not attempt to recreate that information.

Instead, we'll go beyond basic Agile theory to look at successful Scrum Master strategies with real-world examples.

What this book is

An Agile team might have the world's best Product Owner, but without a good Scrum Master, the team won't be able to grow and improve like they might.

Agile training programs focus on Agile theory, and people thrust into the Scrum Master role don't get the education required to succeed. Don't hope for success, read this book and go and get it.

I've spent decades working as, training and working alongside Scrum Masters, both mediocre and exceptional. This book is the culmination of that experience. I will explain the Scrum Master role and how to succeed. I'll also show you what development teams, executives, sales, and the business side of an organization need from you, the Scrum Master, to be successful.

This book provides practical details on how successful Scrum Masters approach essential day-to-day tasks of the job. It teaches how to deliver amazing Agile ceremonies that will inspire your teams.

You will learn to deal with impossible requests, the ones you can't realistically deliver but have been asked to.

I will provide detailed information about goals, strategies, and pitfalls to avoid for every Agile ceremony and primary interaction Scrum Masters will face.

Let's get started!

The Role of a Kick-ass Scrum Master

If you want to slide by doing a so-so job then, by all means, you can describe your role as "The Keeper of Agile Ceremonies." But if you aspire for more (and I'll make the case in this book for doing more), you'll need to broaden the definition of the role. I keep a healthy work/life balance, but I'm still a kick-butt, take names, Rock Star Senior Scrum Master. I like being a full cut above most of my peers because being a rock star has its privileges, doesn't it? When you're a rock star you get better pay, better bonuses, better support from leadership, more choices when it comes to the teams that you work with, and more influence to change the things that aren't working. I wrote this book to help you become a Rock Star Scrum Master.

To be a rock star, you're not just concerned with going through the motions with Agile Ceremonies, but you're looking at big picture results. If you're just looking to complete Agile Ceremonies, you can check that box and achieve what you're trying to do on a failing and unproductive team with low morale and horrible retention. Checking a box on a team that's failing isn't a win, it's just a cherry on top of a big failure. But wouldn't it be depressing to head into work every day to punch time with a failing team? Who wants that? Life is too short, so let's do this right.

Your mission is to "Make the team and everyone on it successful." I would recommend coming up with your own phrasing, but that's the top-level mission, and everything that you do needs to revolve around getting that done. This is a critical change. You're defining the success that you need and letting the strategy come along afterward.

To do the job well, you first need to understand that you have multiple customers, and every customer you have is also someone you are collaborating with. Your first and largest is, of course, your team.

Key Take-aways

- To be a Rock Star Scrum Master, you need to go beyond just supporting Agile Ceremonies.

- Rock Star Scrum Masters get better raises, are more secure during layoffs, and get the choice assignments (you want to be a rock star).

- Your primary role for your job is to make your team and everyone on it successful.

Lead From Inside the Boat

The Scrum Master role is unique among the different approaches used for technology projects because, unlike the traditional Program Manager role, the SM is imbedded directly in the trenches with the team. Even though you are there with the team, there are still two choices you can make about how you see your role and how you're going to go about it. Being in the same meetings with the team doesn't mean that you are part of the team. So, the question is, should the Scrum Master be an outside advisor, or should they be part of the team?

I will give you a fair warning that many people in the Agile community would make a different choice than what I will suggest. I believe that many of them have spent much of their early careers working as Program Managers, having never been part of a development team, and thus don't truly understand the importance of this decision. The choice you make will set the tone of the relationship with the team in every way going forward, so consider it carefully.

Option one would be one modeled more after the traditional Program Manager where you are there to advise and get work done but you're ultimately more of a representative from management than a member of the team. I live in Seattle, and as I commute across a long floating bridge to Seattle from the suburbs, I watch the University of Washington's crews rowing their boats alongside the bridge. These are some of the world's top crews, and they regularly win Worlds. When you see them, there is always a small speed boat following along behind them. This is the model that you'd be doing with this option. When you're outside the boat, you're never going to have the same pull or the same influence that you could have because you're not on the team, you're not part of the team, and everyone knows that you're not going to be losing or winning any races. It also means that if the team makes a severe mistake and ends up flipping the boat or doesn't complete the Sprint objectives, you're not going to be as affected, and you and the team would know that.

Because you're outside the boat, you can watch, but the team won't fully open up to you when your success and motives aren't directly connected to theirs. You're just the advisor, so you can step back and deflect blame for any of the team's failures, and that distance will impact your relationship with your team. You'll never really know how things are working within the team, you'll never know the little adjustments and collaborations that go on to make it all work, and you'll never get team members to fully open up to you because there is that barrier between you and the team.

Option two is where you fully commit to the success of the team. You cross the line from being an outside advisor to a trusted teammate. Unlike other team members who will contribute in authentic, tangible ways (lines of code, automation tests, bugs), your contributions are much more nebulous, so it's going to be that much more important that the team understands that you are fully committed to their success, you are committed to the team, and that you're going to do whatever you can to help the team succeed. When the team understands that you're a fully committed member, the influence, trust, and openness possible far exceeds what you could ever hope to have outside the boat. You can have a good team, but you're never going to have a great team with significant impacts and amazing results until you strap in and join them in the boat, because only then do they know that you're committed. Every great crew needs a good coxswain.

Key Take-aways

- No one has as much influence on a team as the people on the team do.

- To have the most influence with a team you need to be part of the team.

- To get the best communication and transparency, you need to be inside the team.

- The team needs to understand what your role is and how you contribute.

PART 1 – SETTING YOUR TEAM UP FOR SUCCESS

What Is the Standup (Also Called the Daily Scrum) For?

Yes, the standard model that most beginning Scrum Masters learn of "What I did, What I will be doing, Any help needed?" works, but you're leaving a lot of value on the table if you're trying to keep your Standups as short and sparse as possible. Of course, you don't want a Standup with people droning on and on, but you will want to go beyond the basics to do the best that you can for your team.

A good Standup works on a couple of different levels, so the first step is to change the mindset of yourself and your team. The first significant change that I like to make is that I change the name. Yes, the methodology is called Scrum (a Rugby term), but do you want your team to lock arms and try to throw an oblong ball out of a pack of sweaty guys every morning? Probably not!

If you gave the word to 10 different people, you'd get at least six different definitions, and you want everyone to be on the same page as to what it is and what it's used for. If you're a purist, then by all means, keep calling it Scrum, but I like to use "The Daily Planning" because that is what you're hoping to do. Here are the overlapping goals of a good Standup:

1. Find the best ways around the broken builds, tools that won't work, information that you don't have, or anything else that prevents your team from making the best progress possible.

2. Help your team members understand what other people are working on so they can understand how their efforts will fit into a working whole.

3. Provide visibility so other team members can make suggestions, provide help, and collaborate. "I ran into that problem last year, and I fixed it by…"

4. Identify adjustments that would allow the team to be more productive (an hour from someone who knows how something works will save six of someone who doesn't).

5. Over time, establish who is good at what and build the background knowledge of the less experienced team members.

Guidelines for a Good Standup (Daily Planning)

1. Coming out of the meeting with a good plan for the team to be productive is more important than keeping a set timeline.

2. The Standup should be focused and driven by the team, and the team should do most of the talking (not the Scrum Master or PO).

3. Don't discourage sharing if it's gone on too long. Just acknowledge its importance, thank the sharer, and add it to the post-Standup After Party.

4. The parking lot is something you get stuck in, an "After Party" is a better name for the post-Standup meeting.

5. The After Party can be as crucial as the Standup, so stick around and support it.

6. Be patient with team members if they veer off topic a little when they provide status, they will often refocus and finish their thoughts.

7. Acknowledge and thank useful behavior like collaborating, helping teammates, and sharing valuable information, because those are the reasons you have Standup.

8. Provide a clear beginning and ending to your Standup.

9. Provide a clear beginning for the After Party.

10. Keep a record of the After Party items so they don't get lost.

11. Set up the meeting so that if you can't be there, it can continue without you. (Set up Zoom to allow this.)

12. Set it up so that if you must leave after Standup, the After Party can continue as long as it needs without you.

Key Take-aways

- Don't try to limit Standups to a set time frame. The quality is more important.

- Detailing what team members are working on provides a way for less experienced people to learn.

- Sharing details about what team members are working on lets people understand how their work fits into the whole team's effort.

- Sharing information about the challenges or problems they are facing creates a chance for the team to provide solutions, which will save a lot of time and effort.

Agile Ceremonies

Agile Ceremonies are formal meetings held during critical times in the Agile lifecycle, with set deliverables designed to ensure a positive and productive outcome.

The Scrum Master has a central and essential role during the ceremonies that make up the Agile lifecycle. These ceremonies are critical to the entire team and performing well in these events will ensure the team's best chance at success.

The Agile Ceremonies are significant milestones in the Agile process and are just as important to the progress and eventual completion of a project as the daily coding and testing is. The following is a detailed guide that explains how successful Scrum Masters approach each ceremony, how to execute the goals for each event, and how to avoid common pitfalls.

The Ceremony/Situation: Daily Standup, also known as the Daily Scrum or Scrum.

Agile Description: A short daily planning session with the Scrum Master, Product Owner, and the Scrum team brings visibility to the team's progress and ensures that anyone encountering problems can get help. Participants usually touch on what they did yesterday, what they plan to do today, and if they are running into any issues or need information.

The SM's Primary Role: A facilitator, Agile teacher/guide, and a resource for the team.

The SM's Secondary Role: To act as backup for the Product Owner.

The Daily Standup is the most critical ceremony and the most challenging for the Scrum Master. During the Standup, a good Scrum Master will help provide structure without taking over the meeting like a manager or conductor would. If you are just starting out, the Daily Standup might seem daunting. To get better at the Standup, focus on learning to add one improvement at a time and you will make steady progress. A good Standup takes time and craft to establish, but it's a critical skill worth investing in.

Your Goals:

1. Ensure your team can be as productive as possible for the day by ensuring that:
   a. The team talks through and comes up with mitigations for issues or challenges that might slow them down.
   b. No one is stuck or needs help.
   c. Team members understand their priorities and the Acceptance Criteria of the User Stories they are working on.

2. Team cohesion is improved or at least preserved.

3. Newly emerging risks are identified and mitigated if possible.

4. Impacts to the team's work are understood.

5. There is a space where team members can provide and receive help.

How to Execute:

1. Get to the meeting space (real or virtual) early to ensure that everything will be set up on time.

2. Encourage social contact before the meeting starts.

3. Listen carefully to the team.

4. Give the team space to carry on most of the conversation.

5. When necessary, ask questions to understand the progress of the work being undertaken.

6. Listen for engineers who might be stuck and need help.

7. Give specifics around how you will attempt to get engineers the assistance or information when they need it.

8. Acknowledge the value of any conversations that might take too long and move them until after the Standup.

9. Listen for and be ready to volunteer to take items that might be slowing down the team.

10. Listen for and be ready to volunteer for any coordination that might be needed with other teams. (ex. "I can reach out to the Tiger Team and get that information.")

11. Provide reminders of where the team is in the Sprint cycle. (ex. "OK, today marks the halfway point of Sprint 3.")

12. Try to insert one short Agile teaching moment each Standup. (ex. "Focusing on one User Story will help you finish it quicker than trying to make progress on multiple Stories.")

13. If difficult technical discussions or decisions need to be made, or if coordination or decisions with another team needs to be made, then suggest and be ready to set up a short meeting.

Video/Zoom Considerations:

1. Use and encourage your team to use video to deepen relationships and open communication.

2. Look for body language that is tentative or doesn't match what's being said.

3. When possible, socialize before and after the meeting to build team cohesion.

What to Avoid:

1. Don't keep engineers to the short format of "What I did yesterday, What I'll do today, Help needed." This format is too short and doesn't provide the needed visibility. You and everyone else on the team need to get the details to understand what is happening.

2. Don't dominate the conversation; this is the team's ceremony.

3. Let the team and the PO know if you won't be able to attend.

4. Don't assign work to engineers (or let the PO assign work). How work gets done is *owned* by the team. Even if it's difficult, they need to figure out how to do it.

What Can Go Wrong:

1. The Product Owner may try to take a management role in the meeting by pushing team members for progress, assigning work, dictating product designs, and micromanaging the team.

   Talk to the Product Owner one-on-one to explain that:
   a. The team needs to have ownership over who works on which items and how items are designed.
   b. Teams with ownership are much more motivated and engaged.

2. After talking about what they are working on, some team members may move into a more detailed discussion such as how to solve a problem that they are having.

    a. Be patient for a minute to allow the person a chance to complete their idea.

    b. Cut in and thank the person for bringing up the topic they are discussing "because it's important," then tell the team that "In order to give this topic the time that it needs, we will put this item on the back burner and discuss it once Standup is over."

    c. Write down the topic.

    d. Once standup is over, re-introduce the topic and ask the person who brought it up to refresh everyone on the topic.

The Ceremony/Situation: Backlog Grooming.

Agile Description: These are working sessions to improve and socialize User Stories which are likely to be worked on soon. Use these sessions to refresh old, out-of-date User Stories that were groomed too long ago (if older than a Sprint-and-a-half, consider refreshing them).

The SM's Primary Role: This meeting is a presentation of the User Stories, features, and Epics expected in the upcoming Sprint by the Product Owner. The Scrum Master is there in the background to make sure that the meeting goes smoothly and to facilitate as needed.

The SM's Secondary Role: To act as the Product Owner's backup if they cannot attend.

Your Goals:

1. For the Scrum team to discuss each User Story, feature, and Epic with the Product Owner until the team truly understands and agrees on the content and wording.

2. For the Scrum team to estimate most or all the User Stories discussed during the meeting as a group (without the vote being dominated by any one person).

3. For the Product Owner to understand what information, clarification, or modifications the team needs to refine any Stories, features, or Epics.

4. Introduce the Epics and User Stories that make up the Epics with everyone on the team.

5. Ensure the team has identified any additional User Stories needed.

6. Improve the User Stories by:
   a. Ensuring the team understands the User Story or Epic's goal based on the title.
   b. Editing the Acceptance Criteria for clarity based on team feedback.
   c. Capturing any assumptions or edits the team suggests.

7. Record and communicate any new risks that were identified.

How to Execute:

1. Support holding these meetings weekly to keep them from becoming too long.

2. Let the Product Owner facilitate and run the meeting and help as needed.

3. Remind the Product Owner to send out the Epics and Stories beforehand so people have time to review them.

4. Spot check the Epics, features, and User Stories to ensure that they have been written well. Provide suggestions and examples of why well-written User Stories are important and save the team time and effort.

5. Ensure that the Product Owner is listening to the Scrum team and is open to the team's suggestions and edits.

Video/Zoom Considerations:

1. Using video during Grooming Sessions allows team members and the Product Owner to see body language that might show someone not being as confident as the statements that they are making. Use video to allow for better communication.

2. Ask if the team understands each Epic or Story and watch for body language indicating that someone might not.

What to Avoid:

1. Canceling Grooming Sessions. The need can pile up quickly, causing long "growling" sessions.

2. Grooming User Stories that aren't ready. These are Stories that don't have customer types, customer value statements, or defined customer value.

What Can Go Wrong:

1. It may be discovered that key features aren't defined well enough, or the design proposed by architects is

flawed and won't work. Don't try to hide or sugar-coat these challenges. Take the following steps (in order):

a. Identify the people and information you need to clear up this challenge.

b. Help, if necessary, to set up meetings to go over and do the work.

c. Help to inform your leadership about the challenge and how the PO/Scrum team plan to solve it.

The Ceremony/Situation: Sprint Review, also known as a Sprint Demo.

Agile Description: Agile teams demonstrate to stakeholders what they have completed during the Sprint.

The SM's Primary Role: Act as a backup for the Product Owner and represent the team if needed.

The SM's Secondary Role: Stand in for the PO.

Your Goals:

1. Support a good presentation that will:
   a. Demonstrate how your team delivered on its goals.
   b. Show the value your team was able to create.
   c. Show off working code and functionality as close to a production-like environment as possible.
   d. Detail any risks that might endanger your current or future commitments and what mitigations are available.
   e. Talk about the goals for the next Sprint.
   f. Instill confidence in the Scrum team's capabilities and technical innovation.
   g. Highlight the hard work and technical competence of your team.

2. Support the Product Owner to:
   a. Highlight the value that was created.

b. Demonstrate an understanding of the current product priorities.

c. Show their understanding about what the team should work on next.

3. Boost the morale, confidence, and motivation of the team by giving them main stage recognition for all the hard work that they have completed.

How to Execute:

1. With an experienced and prepared Product Owner, the Scrum Master won't have much to do at the Sprint Demonstration except to be available as the Product Owner's backup. Until the Product Owner fully understands their role in this ceremony, the Scrum Master will need to be available to help support.

2. Encourage the Product Owner to use a PowerPoint/presentation template. A good template will look professional and require little time to update for future Sprint Demonstrations.

3. Ensure that the Product Owner and the team have talked about any newly emerging risks prior to the Demo. This not only helps identify risks that the PO may be unaware of, but it also clearly communicates to the team that you hear their concerns and respect their opinions.

4. Get your team to think about what they want to demo several days beforehand.

5. Suggest having the engineers prepare and conduct a dry run of the Demo the day before the presentation. This is critical to being able to give a smooth and polished presentation.

6. Include ways to mitigate risks in your presentation (if risks can be mitigated).

7. Suggest that the PO send out an email after the Demo highlighting all the content and team accomplishments.

Video/Zoom Considerations:

1. Suggest that the PO:
   a. Wear professional clothes.
   b. Talk slower and put emphasis on the important points.
   c. Practice beforehand.
   d. Have notes to work from so you can give a good presentation via video.

What to Avoid:

1. Suggest that the team avoids showing work or code that isn't easily demonstrable. Instead, offer a few screenshots of code and talk through what the team

accomplished; tell the stakeholders how it will improve the customer's experience.

2. Don't try to take the credit for the team's accomplishments. Giving credit where it's due helps motivate your team and shows Scrum Master leadership. Deflect compliments and praise from yourself back to the team.

3. As much as you deflect praise and give it to the team, absorb any criticism. If there is a criticism of the product shown in the Demo, own it.

4. Make sure that the PO doesn't skip talking about risks. Visibility into what could go wrong is essential in the success of the team. It's important for the leadership to know that the team is thinking about and working to mitigate risk.

5. Encourage the PO to showcase what your team will be working on in the next Sprint. A good presentation will highlight the importance and value of what your team will deliver at the next Demo.

What Can Go Wrong:

1. If the code or functionality crashes during the presentation, keep calm and continue speaking while the issue is resolved. If the issue can't be resolved,

explain the work, value, and functionality that is complete.

2. Audio and presentation problems are common issues encountered during demonstrations. Conduct a quick practice with a small audience to ensure that your presentation can be seen and heard.

The Ceremony/Situation: Sprint Planning, also known as Sprint Kickoff.

Agile Description: A ceremony to review and agree upon the Sprint goals, the priorities for the team, and the User Stories to be worked on during the Sprint.

The SM's Primary Role: Facilitate the meeting by introducing the meeting goals and the team's capacity, as well as cuing the Product Owner on the information that is needed.

The SM's Secondary Role: Stand in for the PO.

Your Goals:

1. Get the Product Owner and the team together for the meeting.

2. Ensure that the PO has defined a Sprint goal and set the priorities so that the team knows what is important.

3. Get the information from the PO and be there at the meeting to present it if the PO can't make it to the meeting.

4. Make sure that the team has been able to review and fully understands all the User Stories that they will be working on for the Sprint.

5. Go over any edits or changes to User Stories, features, or Epics.

6. Balance the team's capacity with how many Story points there are to work on and adjust if there is too much work to do. This might include postponing or splitting some User Stories.

7. Leave the meeting with every team member having a User Story to work on.

How to Execute:

1. Encourage the PO to send out a prioritized list of the User Stories before the meeting.

2. Check with the team and PO to see if the User Stories being worked on in the Sprint are ready and any needed edits have been completed.

3. Ask the PO to define a specific Sprint goal if there isn't one, and make sure that the goal has been recorded.

4. If there isn't a Sprint goal, ask the PO what the most critical thing to be accomplished is.

Video/Zoom Considerations:

1. Encourage the team to meet using video.

2. Calm any nerves if last minute changes are required.

3. Record the Sprint goals someplace where they can be revisited.

What to Avoid:

1. Not having needed edits done before the meeting.

2. Not knowing what the top priority is.

What Can Go Wrong:

1. If there are Stories that aren't complete and ready to be worked on, focus on getting enough work to engage everyone on the team and finish prepping the Stories as soon as possible. Don't start work on incomplete Stories.

2. New information may come up which invalidates User Stories or features your team was about to work on. If this happens, acknowledge the new situation to the team and take the steps identified above.

The Ceremony/Situation: Sprint Retrospective.

Agile Description: The Sprint Team gets together to understand what is going well and should be preserved, areas that could be improved, and which actions the team will take to improve one or two of the most important challenges.

The SM's Primary Role: To facilitate the Retrospective and help the team identify ways that they can improve.

The SM's Secondary Role: N/A.

Your Goals:

1. Make sure that everyone on the team feels listened to and has a chance to voice their ideas and concerns.

2. Leave the team feeling good about the things that they are doing well.

3. Make sure that what to focus on is a team decision and not a decision dominated by a vocal or senior person.

4. Leave the meeting with a plan for how to improve one of the top issues identified by the team and who will be doing that, or at least who will own looking into how to make improvements.

5. Keep the meeting focused on the issues and away from finger-pointing or personal attacks.

6. Keep the team's morale and team cohesion intact.

7. Provide a judgement-free space where it is safe for anyone to bring up ideas or concerns.

How to Execute:

1. Before the meeting, connect with team members who will not be able to attend the meeting and get their feedback about what is and isn't working so that you can represent those ideas during the Retrospective.

2. Start the meeting with an introduction to the Retrospective, the goal of the meeting, and the ground rules.

3. At the meeting, review the focus of the previous Retrospective and discuss how much the area has improved.

4. Act as the timekeeper and set a time for team members to list things that are going well and things that might be changed in order to provide a better result.

5. Read through each list and group together duplicate items.

6. Read each item out loud and allow space for team members to elaborate and ask/answer questions so that everyone understands each item.

7. Explain how dot voting works, act as a timekeeper, and give everyone enough time to dot vote for things that are going well and things that might be improved.

8. Score each list and read through the top items.

9. Explain the importance of keeping focused on the things that are going well so that those strengths don't get lost.

10. Explain that, when making an improvement, it works best to select one item and really focus on making changes there. Select the top vote item from the "could be improved" list. If the item selected could be improved by the team, then ask the team how they want to improve that area. If the item is something outside of the team, then take it on and move onto the first item that the team can work internally.

11. Once the team has come up with ways to improve an area, take notes about how and who will be making the change.

12. Thank everyone for participating, explain where the notes will be kept, and close the meeting.

13. Post the notes to the team page or to the daily Standup meeting (anywhere it will get seen).

One of the biggest challenges during a Retrospective is the need to allow everyone to voice a concern or an idea and ensure that a vocal minority doesn't dominate what the team ends up working on. The Lean Coffee format works well at covering both needs while keeping team members engaged.

Video/Zoom Considerations:

1. If this meeting is being done via Zoom, the team will need to have a way to list ideas and vote on them. Trello or other online tools that allow for dot voting are good options. A spreadsheet in Office 365 isn't the most glamorous solution, but it will work in a pinch.

What to Avoid:

1. Letting anyone dominate the conversation.

2. Allowing the conversation to drift into finger-pointing.

3. Trying to direct the team about what they should focus on.

4. Telling the team how to fix something.

5. Skipping the introduction or meeting wrap-up.

6. Not publishing the notes.

What Can Go Wrong:

1. Senior members, or strong-willed team members, can get locked onto an issue that other team members don't see as a major concern.

    Solution: Give the person enough time to describe the issue (to ensure that everyone understands the issue and to give the person the chance to be heard), and then move onto the next item. The dot voting by the team will identify the items that the team truly believes are important.

2. Team members may get into conflict, which might include finger-pointing or other types of blaming.

    Solution: The Scrum Master should pause the conversation, remind everyone that they are on the same team and that everyone wants better results, and reiterate the ground rules to help focus this energy towards identifying a productive solution.

3. There may be no one volunteering to take on improvement items or researching potential solutions.

    Solution: Ask for volunteers, sit back, and wait. If the Scrum Master is patient and willing to let the silence drag on, volunteers will almost always step forward.

Key Take-aways

- Teams that hold quality Agile Ceremonies improve and thrive.

- Understanding the goal and role of the Scrum Master in every ceremony is essential.

- The Product Owner or Dev Manager are natural backups if the Scrum Master can't make it to a ceremony.

- During some Agile Ceremonies, the Product Owner has a more prominent role than the Scrum Master. Support them without stepping on their toes.

- In all the ceremonies it's important to know the goals and strategies to succeed.

Good Retrospectives

If you were to pick a single moment in time where you can make or break your team, the one hour of the Retrospective would be the place. If you can invest and hold a good Retro, you'll improve your team, empower them to take ownership of how the team runs, and take an evolutionary step forward. A Retrospective is also a place where the team is looking deeply at how things are working. If it turns into finger-pointing or blame exercises, it can do extensive damage to team morale and the Team Capital that holds everything together. Here is what you'll want to do to hold a productive Retrospective and sidestep the pitfalls inherent with the process.

Retrospectives can be stressful, so you'll want to try to create a calming and relaxed atmosphere. Only families or people who care about each other eat together (which is why salespeople have a budget to take people out for meals), so the first step is to bring in some food – preferably something tasty and healthy like cut-up fruit.

If you have extra time or budget, decadence is always appreciated (think something chocolate), but if you can only bring one thing, go with the fruit. The second step is to dim the lights a little so there is a cozier feel in the room.

You want to start your Retrospective off with the goal of the meeting. To ensure that it's not going to become personal, you want to establish the ground rules for the Retrospective by reading through them and having a printed or written copy in the space.

Ground Rules

1. This Retrospective is about us getting better, and it will take everyone's participation to do that.

2. We will talk about what happened but stay away from who did what (no names).

3. What is said here stays here.

4. We will be respectful and treat each other with kindness.

Once everyone is in the room, and you've walked everyone through the ground rules, you'll want to kick off the meeting with the Retrospective Introduction. Here is the Introduction that I use, and this would be a good starting point for the one that you use, but I suggest customizing so that you have full ownership and are comfortable with what you are saying.

Retrospective Introduction: Team management has entrusted us with the ownership and responsibility to decide for ourselves how best we can work together. It is very unusual for a company to empower a team to make these kinds of decisions without a manager or executive peering over our shoulders, but we have shown that we are up to that task. We can look and change the way that we work at any time, but this time has been reserved for making sure that nothing slips by our notice that should be changed. There are things that we are doing now that are working well for us, and we must recognize those things and continue to do them. There may be things that used to work very well for us, but due to changes in where we are in the product lifecycle or in how we're working together as a team, may no longer be working as well as they used to. We all have our own opinions, and I have mine, but mine isn't any more important or less important than your own. This is our team, and we decide how it works, so this is where you have a direct voice in that future.

Listening and Side-stepping Unproductive Topics: One of the more significant challenges with Retrospectives is that there are always people on the team that have hot button topics that they would like to see addressed. Everyone has hot button topics, but sometimes these are things that should be addressed, sometimes these are topics that would be hard or impossible to fix, and sometimes there are problems that may not be causing as much trouble for the team as one person thinks they are. So, it's critical that everyone on the team feels free to bring up any issue that they like and has a chance to talk through the problem and be heard, but these may not be things that the team should act on. The best way to accomplish both goals is to conduct voting (such as dot voting) for the topics that are brought up. Voting means that there isn't a single person who is deciding what is most important for the team and ensures that the whole team will be choosing.

When Problems are Outside of the Team's Control: It's not uncommon for the team to be encountering serious problems that are outside of their control. Issues like these can provide a unique challenge and be a source of frustration for the team because these problems are often larger company-wide problems and can't be fixed by the team directly.

Don't stop the Retrospective in its tracks, but at the first natural opportunity acknowledge the unique challenge that this type of issue poses because it is outside of the team's control, and then acknowledge the impacts that the issue is having on the team. As the two people who represent the team and spend a lot of time interacting with groups and people outside of the team, these types of issues should be owned by the Scrum Master and Product Owner. If you already know how you plan to address the issue, then go ahead and tell the team that you own the issue and detail your plan to resolve it. If you don't have a strong and coherent plan, then just take ownership of the issue, and get back to the team once you have a plan on how to address the problem.

Example: I was working with a team that was having challenges during quarterly cross-team planning. After preparing for these sessions, the team would often get last minute changes on the day of the event. Leadership would remove new product features and work that the team had broken down and understood and add new work that the team didn't understand and wasn't ready for. These new requests created chaos as the team had to drop active planning to break down and try to understand the new requests. Because this problem was outside of the team, as the Scrum Master I took ownership of the problem and approached the leadership of the program with details about how the issue slowed down and degraded the quality of the team's planning. A meeting was added two weeks prior to the quarterly planning event where the Program Managers and Architects coordinated and delivered a list of the features and Epics that each team would be working on. This dramatically reduced the number of last-minute changes that the team encountered.

Focus, Focus, Focus: When a professional athlete wants to improve, they get the best results when they just concentrate on one thing at a time and work hard to get that one thing right. Scrum teams are the same way. To make robust and meaningful progress, you want to leave the Retrospective with just one or two things to work on, plus owners for those items. Once the Retrospective is over, you want to go and post the findings in a team space or, if you are working remotely, into the daily Standup invite so that it's visible for the entire team.

Key Take-aways

- Retrospectives can be great, or a significant blow to your team, so take them seriously.

- Do what you can to set up a comfortable low-stress environment (bring healthy snacks if you can).

- Introduce the Retrospective to firm up the common goals of the team.

- Vote and take only the top one or two things to work on.

- If the issue can't be fixed, acknowledge it and move on.

- If issues are outside of the direct control of the team, then the Scrum Master and/or the Product Owner should own them.

- Don't assign the team to fix or contemplate problems that are outside of their control.

Public Relations

Building products or software is already hard but doing it with speed and quality is incredibly difficult. Things are going to go wrong, and sometimes it's going to get a little messy, so it's essential to keep up the team's spirit and capture the things that go right. Even if you're on a team that's struggling, there will be things going well, and you want to be able to capture those because it's likely that if you don't, no one else will. That may sound harsh, but it's true. You see it all the time in the corporate world, where people with undersized talent and work ethic latch onto wins (often those that they didn't contribute much to) and wave them around to get outsized applause and kudos. So, while your team is doing significant work and making progress, why wouldn't you want to get proper credit for what they are accomplishing? It's a classic reoccurring theme that people, and especially developers, don't get the recognition that they deserve. All too often in their careers, they watch Program Managers and other people stand up to get the credit, applause, and bonuses in company meetings when the dev teams were the ones that made it happen. When you've put in that much work, and you see someone else get credit, it stings no matter who you are.

Consistently, not being recognized is at the top of the list for employees who are unhappy with their workplace, yet it's also (unfortunately) incredibly common. The great news is that it's straightforward to fix and has over four times the pay-off! Let me explain. When your team gets some of the recognition that it has been missing, it pays off on the:

- Personal Level – The person who put in the great work gets recognition for their efforts, giving them a feeling of accomplishment and gratitude towards you for recognizing what they've done.

- Team Level – Your team will see that you're looking for, recognizing, and voicing when they do great things. This will encourage them to want to step up in the future.

- Project Level – The leaders and other teams on your project will see the positive work that your team is doing and will want to step up their work to match the good things that are being done by your team.

- Organizational Level – Budget and headcount are assessed and handed out regularly. Executives and managers want to make sure that resources are being used efficiently and to good effect, and when vital work needs to get done, they want successful teams to take it on. When your team has a consistent, repeatable drumbeat of success, you're going to be at the top of the list for headcount, budget, and exciting work.

- Bonus – Just as your team needs PR, your boss, as well as their boss, needs PR too. Especially at the executive level, the role is hugely political, and to do well you need to have a steady, regular stream of successes. The wins for your team turn into wins for your boss and your boss's boss (and it doesn't hurt to earn a few brownie points and recognition).

- Bonus Bonus – As the designated servant-leader for your team, when your team is looking good, working well, kicking butt, and taking names, then you will look great. You look much better than if you were trying to talk up your own accomplishments.

Finding places that are working well doesn't do anyone much good if you don't point them out. Pointing them out helps the team notice them, gives a boost to the people who are responsible, helps model positive behavior, and lets everyone know that you're paying attention and may share if you see something positive that they are doing in the future. For struggling teams, this is especially important, because if the team is struggling, everyone can feel it, and it's never fun to work on teams that are flailing. Feeding your team is like feeding a pet. You could fill up their dish with a week's worth of food on Monday morning, but if you do it all at once, there isn't going to be anything left by Friday. It's best to provide small wins regularly. Of course, these need to be genuine. The last thing you want to do is lose credibility by trying to take a win with something that everyone knows wasn't a win.

OK, so maybe I've convinced you that you need some PR. But, getting back to reality, how are you going to accomplish that? You may not have the money to hire a team, Development Managers have lots of meetings, and Product Owners are (hopefully) meeting with the business or with customers. The good news is that you and your PO can do it, and it really won't take very much time at all. As the Scrum Master, it's up to you to notice the people on the team getting the work done. No one knows the wins and inside story as you do, and there is always the time, money, and energy for what's essential. The hard part is getting into the habit of paying attention and calling out the good things that happen. Some of this work will only need to be done once, so it works best to knock those out first. From there it's an easy downhill slide.

The most important thing with PR is that it isn't a one-time thing, the praise is justified (not overboard), and it's done in a regular repeating cadence. For it to stick, it needs to be a typical, consistent message, not something piled on heavily occasionally. Below is how to set it up, and the best part is that you don't need to spend much time doing it. If you focus, this should take you less than two hours, and once it's set up, it's easy and on autopilot, so invest in yourself and do this.

One-time Items

- Set up a reoccurring reminder/meeting for yourself and your PO to meet to list out and find wins before the retrospective.

- Set up a reoccurring reminder/meeting for yourself to find something positive twice a week.

- Set up a reoccurring Demo meeting (if you don't already have one).

- Set up a reoccurring reminder for your PO to send out the Demo Notes. Create a Demo Note email template with a catchy title and attach it to the reminder.

- Create a PowerPoint template for Demos. (What if we're not doing Demos? What if we do Demos, and no one shows up? What if we do Demos, but no one uses a PowerPoint? Well…. If any of these are correct, then congratulations, you're about to look like a rock star with these templates!) The template should be short and simple, so it's easy to fill out. Your Demo PowerPoint template should include:

  o The team's focus and the business objectives for the Sprint.
  o Accomplishments for the team.
  o Focus and objectives for the next Sprint.
  o Risks and dependencies for the next Sprint.
  o Team members.
  o Who is making the Sprint and what it's showing.

The Importance of the Sprint Demo: The Demonstration of working software is often overlooked, and some organizations don't hold them regularly, but getting rid of them would be a mistake that you don't want to make. The Demo is where the work that your team has done gets to shine, and your Product

Owner gets to showcase their command and business savvy by laying out the goals achieved and the goals and strategy for the upcoming Sprint. For visibility and Public Relations, this event is as good as it gets.

I know that for some organizations, it may be challenging to get your stakeholders to come consistently. Still, as the Scrum Master, it's worth sitting down with your stakeholders to explain the importance of the ceremony and how it's in their best interest to come. When stakeholders consistently come, the team and everyone knows that the software needs to be in a decent working state, and the quality of what the team is producing will go up. Even if the stakeholders don't consistently come, you can still record the sessions and send those along with a Demo recap email.

The secret is that the essential part of a Demo isn't at the Demo, but it's sending out the notes afterward. Most people don't bother to send out notes, or if they do, they aren't consistent, which is missing a big win. Your boss and your boss's boss need to demonstrate successes and value creation. If your boss has an inbox or saved folder full of these emails (Team Foo's Big Wins Sprint 1, 2, 3, 4...) they have a regular useful drumbeat of success that they can save, forward, refer to, mine, recycle, share, and wave around. In tech, things move fast, and if they don't have a copy handy, they won't be able to remember what happened at the Demo if it was more than a few days ago. The smart thing to do is to create a single template for your Demo emails, get it attached to a reminder, and then the day after the

Demo, you can take five minutes to cut and paste and send out the Demo recap email. You want to use the same title and the same format every time, so everyone gets used to seeing it.

Key Take-aways

- You and your Product Owner need to become the Public Relations for the team.

- Public Relations doesn't take up much time once you have a few templates to use.

- A regular steady beat of PR is better than having it very loud occasionally.

- Having great functionality and a plan for the next Sprint at the Demo is some of the best PR you can get.

- Make sure to email out your Demo notes after every Demo.

Building a Steady Rhythm – Reserving Rooms and Meetings 6-9 Months Out

In technology, product development produces a lot of change. When developing new products or improving existing products, there are lots of hard choices to make. Do you go with existing tools and applications known to have shortcomings, or start from scratch with new, potentially better but unproven technologies that teams are unfamiliar with? Along with work challenges and churn, everyone on your team has a home life that demands attention, scheduling, and time commitments.

To get the best results and a less stressful environment for your team, you want to have a steady, predictable foundation that they can work from. So, you'll want to sit down with the team and talk through when they want to have Standups each day. Mornings are best because they allow everyone to get a refresh, get help/advice to work around any issues they may have encountered the day before, and have some productive face time with the code. With many people on your team, it's going to be hard to please everyone, so you'll need to find something that will work best for most of the team and have accommodations for those it won't work well for. This might include such things as recording the Standup and making it available for team members to watch, as well as having team members who can't attend write up what they would have said during Standup.

Your goal is to have your ceremonies at the same time, the same day of the week, in the same conference room or location (which are close to where the team sits), with the same dial-in (if you need one). This regular cadence provides some assurance, a foundation to work from, and will allow your team members to plan and adjust their schedule to make it work. The longer you can keep the same cadence, the better, because it becomes predictable and something that everyone on your team will remember.

Setting up regular reminders for you to extend your invites is also an excellent way to make sure that you're out in front of the competition.

Key Take-aways

- Having meetings at the same time, in the same rooms, helps simplify things for your team and allows them to focus on the essential work.

- You may need to schedule your meetings far in advance to reserve the same room consistently.

- Set a reminder in Outlook so you will be reminded to extend the meetings when future dates become available.

The Mechanics of Good Meeting Facilitation

I was leading a prominent team at a large international bank, and I had one of the senior programmers who was a crucial leader in the work that was underway. Still, he was consistently late to Standup and would miss the first few minutes of the conversation. This was distracting because we had to either stop and catch him up with what had been said or risk having him out of the loop and not fully participating. After seeing this pattern emerge, I pulled him aside one day into a conference room so I could mention this problem and hopefully get it fixed.

I started with, "Hey John, I've noticed that you've been coming in late to Standup lately, and it's been a distraction for the team because you are critical to the work that we're doing, and we have to stop and catch you up."

He looked a little sheepish, looked down for a moment, then back up to reply "Oh, sorry, I just noticed that we never start on time, and I hate waiting around for things to start."

Well, that was a good kick in the stomach for me. While I hated to admit it, he was right. I had gotten busy and hadn't been starting Standup crisply on time, which means that I also wasn't modeling the behavior for the team that Standups were essential, and we needed everyone there. I'm glad he mentioned it because it was an important lesson for me professionally.

If you want your team to take Standup seriously and value it (as they should), then you're going to need to take it seriously too, which means that you'll need to run good meetings. Here are the things that you're going to want to do.

1. Include agendas for all invites that you send out. These should provide any background needed, what the goal of the meeting is, the topics that will be covered, and any questions that will be central to getting a positive outcome for the meeting.

2. Send out well in advance, so people have time to see, accept, and get ready for the meeting.

3. If you are sending out an invite where the meeting is in less than two days, send out a separate communication.

4. If you are sending out an invite to a group of people outside of your team, then send out a separate communication with the agenda and goal of the meeting.

5. If you are sending out an invite with less than six hours' notice, then contact attendees via chat/Slack to give them a heads-up and provide the agenda/goal of the meeting.

6. Make sure to have dial-in numbers in case someone can't attend in person.

7. Take notes during the meeting.

8. Send out the meeting notes quickly (within two hours), so everyone has a copy while it's still fresh in their minds.

9. If there are technical topics to go over or people that can't make the meeting, record it if you can.

10. Include a link to the recording with the meeting minutes.

11. Be at the meeting three minutes before its scheduled to start. This provides a buffer in case something goes wrong, and you will still have time to recover and be at the meeting on time. This also allows time to get composed before starting the meeting.

12. If you are waiting for last-minute stragglers, then thank the people who did make it on time and engage them until you can start. This is a great time to invest in micro team building by connecting with team members about pets, travel, upcoming vacations, or other non-work items that will help everyone get to know each other better.

13. Start the meeting off by reviewing the agenda and the main topics or questions that you are hoping to attend to.

Key Take-aways

- Run professional, well-organized meetings to get professional, well-organized participation.

- Communicate often and early.

- If your invite is last minute, you need to go the extra mile to get everyone there.

- Include agendas.

- Start on time and get notes out quickly.

PART 2 – MAKING FRIENDS AND ALLIES

Setting up a Healthy Culture

In my second job as a Scrum Master, I started at a large multi-national financial company with a technology development office in Seattle. After getting my laptop and logins, figuring out where the coffee machine was, and setting up my desk, I attended my first Standup with my new team. The team was a well-established senior team critical to the success of major projects and I had never met any of them (even in an interview), so I was a little nervous about what to expect. To my surprise, as I looked around the room, I saw that the team was also a little jumpy and watching my reaction to everything they said with a bit of trepidation. Fortunately, the first week went well, but on Friday, a senior member of the team pulled me aside into a conference room for a word. I didn't know what to expect, so again I was a little nervous about what kind of bombshell I was going to get. After the door was firmly closed, she leaned forward and, with relief and enthusiasm, said, "Things are going so much better than what we expected!" I replied, "Well,

I am just getting started," and she got serious for a second and shared, "Yes, but you're not yelling at us." She explained that the old Scrum Master was regularly yelling and berating the team when even the smallest thing went wrong, and the team had been on pins and needles waiting for my first outburst.

If you think about the kind of culture that yelling sets up, it's easy to see how destructive it can be. If you are going to get yelled at or get someone else yelled at in front of co-workers, are you ever going to feel safe going to your Scrum Master with an issue? I know I certainly wouldn't, and that type of behavior, along with being abusive and certainly destructive over time, would undoubtedly open legal risks as well. No one wants to get into that type of mess, right?

If your team is going to be successful, having a healthy culture is a must. Having a common culture and language is what helps make the team a team, so don't underestimate or leave this to chance. As a Scrum Master, one of the crucial things that you're responsible for is to help create a healthy culture, and the first big step is to model the culture you're hoping to develop. Here are the powerful messages that you'll want to convey regularly and loudly, but more importantly, you'll want to model these.

Message: No one is perfect, and no one has all the answers.

Why This Is Important: To get the team fully engaged, you need them to know that you don't have all the answers, and you don't even have all of the answers when it comes to Agile (even though it's what you're being paid to do). Even if you feel that you have all the answers (because you're just that good), you may not have factored in all the information for a given situation, or there may be an underlying problem that you don't know about. When there is more space for discussion, there is more space for questions to come up and for you to mentor and teach.

Message: We are smarter when everyone participates.

Why This Is Important: Teams typically only have a few senior members, so if the less senior team members hold back and let the senior people take the lead with questions, code, and design ideas, you're only going to be utilizing 20-30% of the team's brainpower. Of course, senior people are a real asset because of the experience and product knowledge that they have. Junior people are also a vital asset because they typically have strong skills in newer technologies/tools, and with a lack of experience they can ask the pointed questions like "Why do we do it that way?" that are excellent for uncovering trade-offs that were made years ago that might no longer make sense. It's going to be critical that everyone participates fully.

Message: This is a team sport.

Why This Is Important: When people on the team start thinking not of their success, but the team's success, then you get smarter decisions and support that doesn't happen until that type of thinking kicks in. For example, I had a team where a senior person was great at getting around build issues and getting people unstuck so they could get back to coding. Over a month, multiple changes in the build process brought people to a standstill five or six times. If Siva had been worried just about his productivity, he could have expressed sympathy and continued on his way. Fortunately for us, Siva was a team player, and with his knack for troubleshooting, he would sit down with team members and fix in 20 minutes what would have taken them two or three hours to do. It also helps them appreciate the work that other people do on the team, and that helps morale.

Message: We help and support each other.

Why This Is Important: When there is a culture of mutual support, then the people who need help will know that people are available and there is an expectation that if asked, someone will help. Having it built into the culture makes it acceptable and maybe even expected for people to ask for help, which will save your team a lot of time when people run into problems. It also encourages collaborating, which leads to better results because each person can provide experience and compensate for the other person's weaknesses.

Message: This is a great team, I'm happy to be heard, and I'm honored to be working with you.

Why This Is Important: When your team chemistry and morale are high, the output, the effort that people are willing to spend, and the enjoyment and dedication to the team all go up significantly. Every team has strengths and weaknesses, but everyone wants to have pride in the work that they do. When you take pride in and appreciate being on the team, so will they.

Key Take-aways

- The team and organization's culture is the foundation upon which everything else is built.

- Without a healthy culture, you don't have a team; you have a group of people where some of them will be pulling in different directions.

- Success requires the full team to be involved.

- Critical decision-making can't be reserved for a few individuals if you want the team to succeed.

- As the Scrum Master, you are the architect of the team's culture. Your primary job is to build and nurture a healthy team culture.

Go Beyond Transactional

Like all developers, my team likes to complain about having too many meetings, but once a quarter, everyone on my team gets an invite for an optional meeting, and people rarely miss it. To make the meeting, some of them must come into work half an hour early, but I never get any complaints. They're just happy to show up. Downstairs (thanks to some architects with a little foresight), there is a standard room with a bit more space, and I set everything that we'll need there. As team members stagger in, each is given a plate with a thin, light crêpe stretched from edge to edge with the smell of vanilla and cinnamon drifting up to caress their noses. After receiving a warm plate, each team member approaches a counter laid out with bowls of fresh organic strawberries, raspberries, Greek Yogurt to add some creaminess to the fruit, and anything else that's tempting and fresh. For the more adventurous, there is a whole other set of bowls for the savory crowd filled with fresh diced bacon, chives, sour cream, mushrooms sautéed with white wine and pepper, and a bowl of freshly caramelized onions in all their glory.

After decorating their crêpes just, the way that they like them, people drift over to a large table, grab a fork, and start to enjoy a warm yummy breakfast. While all of this is going on, I'm busy making the next crêpe, but everything is spread out around me, and I'm just a few feet from the table so I can socialize with people as they drift in without endangering the next crêpe that's turning golden brown on the pan.

There is one significant rule for these gatherings that is non-negotiable, which is that there can be no talk about code, or work, or anything that generally dominate so much of our typical day-to-day interactions. Without that topic hanging over breakfast, team members catch each other up on family lives, where they are going on vacation, and pictures of the kids are shared around.

The great thing about this is that it doesn't take me much work to pull this off. Yes, I bought a plugin deep-sided griddle for $425 years ago, but it lives in a bottom drawer at work where it isn't in the way. Altogether it takes me 15 minutes in the kitchen at home to mix up crêpe batter and put it into a large thermos, plus another 30 to add a few things to my typical shopping list and cut them up while the bacon cooks in the pan. Altogether I probably spend 90 minutes to pull off a hot breakfast for a very appreciative team, so it's easily worth the effort. It's an important event because it sets the tone of our team's relationship. I'm not feeding them breakfast so they will attend a technical training or a company event. It's clearly in appreciation for the work that they do and who they are as people. The friendships and trust that are built during these events make our team one of the company's highest performing and most productive.

If you were to pick a single word, an available title that had the most influence and impact on the world, "Mom" should be at the top of the list. Yes, Mom was there at the beginning, but the critical aspect of Mom is that she is the single most supportive person in most people's lives. No matter what you're doing (great or small), mothers are interested, listening, and cheering for you. Mom wants you to succeed in whatever you're trying.

While most mothers do give a lot of support and interest, they also gain deep loyalty and influence, and most people would do anything for their mothers. Besides Mom, most of the support and interest that people receive is transactional. The developers on your team may get training, but only if it's something they will directly use. People get awards and bonuses, but only if they have done something productive for the company. Don't get me wrong, I think it makes complete sense to reward team members who have gone above and beyond to do something good for the company. It's just that, as a Scrum Master, you want to go beyond transactional support to get all the influence you need.

That may sound interesting to you, but how do you do it? Becoming the unconditional support of your team is not as hard as it sounds, and it truly makes a big difference in the acceptance, trust, and influence you'll have with the team.

Step 1: **Define a Non-transactional Approach:** When I talk about my role and what I'm trying to accomplish, I'm cautious about how I speak about my end goal. If I talk about my purpose as being "here to deliver customer value," "to help the company makes tons of money," or "to help you write great software," then what I'm doing is transactional, so I'm only really helping and supporting the people on the team because I want something. Implied is that if they aren't doing what I want them to, if I think that they aren't delivering, then I'm not going to support them.

Step 2: **Demonstrate a Non-transactional Approach**: The way that I talk to my team is that I tell them that I want them to be successful. Of course, talk is cheap, so along with making that statement, I usually recap with an offer to help with something that isn't work-related. I live and work in Seattle, and I'm one of the rare Seattleites who have lived here all my life (I'm a third generation raising a fourth), so I know a lot about many of the music festivals, food, and fun things to do outdoors in the Seattle area. I tell them that if they need something to do over the weekend to let me know because I've lived here forever, and I know many fun things to do. It's rare, but sometimes I get one person per team who takes me up on it. It doesn't take very much time, but for someone who may be new to the culture and the city, imagine the significance of having someone who is willing to take five minutes to help you find something fun to do with your kids over a three-day weekend. Doing this can be as easy as subscribing to an events email and bringing an item or two to the Friday Standup.

Key Take-aways

- Few people have someone fully committed to their success. Having someone fully committed to your success is rare, valuable, and appreciated.

- Focus on the person you're supporting, not the outcome of gaining influence and loyalty.

- By offering help with trivial non-work items, you show you are there for their success and not just trying to get more work out of them (you're non-transactional).

Building Team Capital

Team Capital is goodwill, trust, and camaraderie that makes the team more than a collection of people and into something that can work as an integrated unit. Neither marriages nor good teams happen overnight, no matter how good the chemistry is. It takes time to establish the trust, communication, and roles that will make it successful. Because Team Capital can't be easily measured on a graph or with a number, it's usually ignored by much of the business world. Still, just like a good marriage, Team Capital produces impressive results. When measuring success, productivity, and happiness, poor Team Capital can be just as devastating as a divorce. In a relationship, it's not like there is a number we can use to assess its health ("We've been at -6.2 for three months, and she won't do anything about it! Time to get a divorce lawyer!"), but that doesn't make it any less real or impactful, so let's take this seriously and acknowledge its importance.

In any relationship, there are going to be disagreements and arguments. The team will need to be able to get beyond these and continue to collaborate, but that isn't a foregone conclusion. When there is plenty of Team Capital in the bank to be able to pay for the withdrawal that arguments or conflicts bring, then the team can get beyond these regular bumps in the road and continue to communicate, trust, and collaborate. As Team Capital gets low, the everyday conflicts will start having an outsized impact on team members' willingness to collaborate, communicate, and trust. In extreme cases, conflicts or deep-rooted distrust can take hold and rip teams apart, and when that happens, almost nothing productive gets done. I know that you don't want to get anywhere close to that type of scenario, so let's invest.

Doing a team-building event once a quarter is good for morale, but it ultimately is only a one-time event, and by the time the next quarter runs around, the small bump in Team Capital has long been spent. The best way to build Capital (and it's effortless to do) is to make lots of small deposits. Frequent deposits tend to compound on each other, and they can build up quickly. There are two foolproof ways to do this.

The first Capital-building method is to eat and cook together. Until very recently, we as humans only ate with our family, so meals are a built-in way to bond and build trust. Cooking together might not be practiced regularly, but it's also an activity that only the closest of friends or family do together, and of course, after you cook, you must eat together, so it's a great bonus activity. The first step is to look at your workspace to see if a lunch table can be added to allow people to sit together for meals. If it can't, then talking with the team and scheduling a regular reoccurring team lunch is a great alternative. I like to make waffles or crêpes for breakfast when the teams have done an excellent job for a release because it's a triple bonus. After all, we're eating together, building the Team Capital, and it shows my and the company's appreciation for the work and effort the team has put out.

Potlucks are always a fantastic way to build morale and Team Capital. Our workforce is very diverse, and many of your team members may be from different countries. I like potlucks because people from different cultures often have a rich, exceptional culinary heritage to bring to the table. When a tasty dish is put in front of co-workers, this difference can be celebrated and appreciated by the full team.

Another great way to build Team Capital is to add simple daily questions that each team member can share. These are easy to include in the daily Standup, add about five minutes, and over time the answers to these questions build up and provide you with a deeper understanding of your co-workers. I like to use Table Topics, which has an inexpensive set of ice breaker cards you can order on Amazon for about $30. Some questions include "What was your favorite meal as a child?" and "What did you want to be when you grew up?". If you use these, just be sure to review the questions beforehand because there are some such as "Who was your first kiss?" which could make some people uncomfortable and should be removed from the deck before you use it. Some perfectly innocent questions might trigger something in someone's past, so it works best to have these questions be optional and something that anyone can pass on without the need for an explanation.

Key Take-aways

- Teams are greater than the sum of the parts.

- Team Capital is what holds a team together.

- Invest early when creating a new team to create Team Capital (and get it going).

- Regular small investments are the best way to make Team Capital.

Listen to Your Team and Team Members

Managers and executives will make decisions and implement policy to try to improve the results of your team, but the people with the best knowledge about what is happening and why are your team members. Even working with them every day, you're not going to have the intimate details that collaborating, working with the tools, using the build and release process, and diving into the code base brings. Managers and executives are going to be even further from the action, but under the old model of work, they are the people empowered with making policies and changes to how the team works.

Getting insights from team members will empower you and them with a deeper understanding of what is happening. This knowledge will allow you to bridge the gap and help the managers and executives to understand what's happening under the hood of your team, allowing them to make better and more informed decisions. Listening to your team is essential, and one of the primary ways you'll want to decide what to focus your energies on.

Before you can listen to your team, you're going to need to nudge them to open up. You might have a few people on your team that are comfortable sharing, but being forward with information isn't the norm, especially in a corporate environment. The first step will be to change how you're going about your job. To get your team's feedback, you're going to need to ask for it a lot more. As I've gotten better at my job, I realized that asking a leading question is powerful because it helps the people on the team to think through and come up with their own answers. The team then has much more buy-in for any solutions and conclusions they come up with alone.

Asking questions is also significant because, as an expert, you're not pretending that you have all the answers (and if you feel like you do, you need to recalibrate yourself). Knowing that you don't have all the answers is decisive for the team because they will realize that they need to think about how things are working and take more active ownership. After all, even as an outside expert, you're not going to solve everything for them. Ultimately, this is a healthy change to make.

The first step is to start asking a lot more open-ended questions in your ceremonies. Phrases like "How is this working?" "What challenges do we have here?" "What's slowing us down?" are all great additions and should be regularly sprinkled into your ceremonies.

The second thing to do is to identify your subject area experts on the team and regularly check in with them on how things are going in their area. On a previous team, I had a senior QA Engineer who was regularly much more candid about problems with the automation program with me than he ever would be in front of the team. You'll want to keep tabs and remember the strengths and weaknesses of the people on the team.

Key Take-aways

- The people with the most knowledge about what is going wrong with the team should work on it.

- The people with the most knowledge about how to fix things on the team should work on that.

- As a whole unit, the team is smarter and more capable than any one person (even you).

- Asking questions is more powerful than giving answers or advice.

- Asking questions gives people ownership on the solution when they come up with the answer themselves.

- It's essential to check in with subject matter experts regularly.

Develop a Thick Skin

For any given job, there are perks and there are the downsides – the things that make it hard and not something that just anyone can do. Being successful as a Scrum Master means having a thick skin and the emotional intelligence to be able to handle having someone mad, annoyed, or grumpy with you without destroying that relationship or modeling behavior that would be destructive for your team.

Let's get real here for a minute, step back, and look at this. Technology is a field where you have a lot of super-smart engineers who aren't used to dealing with people, and they often don't have strong people skills. It's also a stressful, fast-paced environment where the burn rate for a good development team could be six million dollars a year and millions, sometimes billions, are at stake.

Your role is to be the calm at the eye of the storm. You need to be the person to pull things back when the emotions are high and make peace when teammates are arguing over the technical direction of the product, which means you're going to need to have a cool head. When things get tense, here are some of the tools, patterns, and messages you will want to have close by.

Common Cause

Use When: There are disagreements on the team that might be getting personal.

Message: We're not always going to agree, and having different opinions is healthy because we can talk this through, understand the choices better, come to an agreement, or discover a third option that gives us more advantages than either of these choices. As a team, we all want a good outcome, and we're all going to support the direction that we'll eventually take. We just need to talk through the pros and cons of the choices that are in front of us.

A Fresh Look

Use When: The conversation has progressed but has taken a different direction and is no longer working toward the original goal.

Use When: You think that the solution or direction being proposed doesn't make sense or may not be considering major important factors.

Message: Let's step back and review what we know so far...

A New Voice

Use When: The conversation is now stuck and repeating or not making forward progress towards the goal.

Message: Thanks, [Insert Name], that's a great point. Does anyone else have thoughts about what could be happening in this area?

Re-center on Our Goal

Use When: The conversation has switched and is no longer making progress towards the original/most important goal.

Use When: Emotions are running high.

Message: OK, let's review; our goal here is to…

Key Take-aways

- As a Scrum Master, a vital part of the job is to be the cool head in the room.

- When tensions are high, some patterns can help focus emotions and energy away from interpersonal conflict and towards fixing things.

- Conversations can get stuck and going over the same arguments is frustrating for everyone, so use patterns to get unstuck and move on.

Don't Take the Squawk Too Seriously

At the beginning of a six-Sprint quarter last year, I saw that the team had a lot of difficult work ahead of them and it had been a while since I had taken a hard look, so I went back over the entire calendar and looked at every meeting that we had on the books for the team. After close examination, I saw that there were several meetings that no longer held as much meaning as when they were created and one that could be shortened up, reducing a 60-minute meeting (it usually went 40), to half an hour.

I looked over everything else, saw that our meeting schedule was as tight as it could be, felt satisfied that I had kept it to the bare minimum, patted myself on the back, and went off to the next thing. The following week, two new developers joined the team from somewhere else in the company to help handle the increased load, and the week after that, I was in a conference room fielding a question from the Development Manager I partner with. The manager had gotten feedback from some of the developers that, as a team, we had too many meetings, and was there a way we could cut any of them back?

I could have gotten angry with the unnamed developers who were complaining, or I could have tried to squeeze out additional meeting time and leave the team with not enough planning time to be able to prepare our work correctly, but instead of getting angry I fell back to a defensive position and started asking questions. I explained to the Dev Manager that I had already removed all the meetings that I could, but I'd love suggestions on other places where we could cut. He came back the next day with a recommendation from the unnamed developer that we only have Standup twice a week. Of course, that was a bad idea, because while we might save 45 minutes a week (3 x a 15-minute Standup), the team would miss out on the critical coordination and information exchange that helps us get around the broken builds and stuck people that slow down our productivity. I also could have shut down the discussion by telling the Dev Manager that I had already gone over everything, but that would have discouraged the person who was suggesting it in the first place and make it less likely that they would try to think and propose future improvements. So, it's essential to be careful about how you react when you are getting feedback. Getting feedback is critical, and it's vital that people feel like they are being heard.

Even if you disagree, or if you've heard the same concern multiple times, you must listen carefully. Don't cut the person off, ask clarifying questions so that you fully understand the situation (and the person feels like they are being heard), and thank the person for the feedback. Maybe this is something that you've heard before, but you want them to know that you're taking this seriously. Do take this seriously, and appreciate it, because you want them to come back with other ways that things can be improved.

I get it; there are always things that people hate and love to spend time complaining about. For instance, most everyone will complain about slow traffic, the cost of housing in large metropolitan areas, and how much property tax you end up paying each month (and no, I'm not thrilled about any of those things either, but complaining and dwelling on them won't make any of them any better). Developers on technology teams exist because they like to, or at least are good at, wading into complex problems and coding solutions to deliver the value that customers want. Honestly, for most developers, providing for the customer isn't a primary motivation. They just like to make cool things, and the last thing that they want to do is to have someone take up their valuable face time with the code with meetings.

Feedback is essential, and while you must be listening to the team and individual concerns, you also can't let the grumblings of a few individuals dominate the decisions for the entire team. This means that when you are listening to feedback, suggestions, and people pointing things out, make sure that you are also getting second and third opinions from other people and balancing the needs of a few versus the needs of the team. It goes doubly so when it comes to meetings. Once you have 10-13 people in each meeting, you're almost guaranteed that someone will have a conflict, so you're just going to need to determine who the critical members are and try to find the time when the most required and nice-to-have people can make it.

The challenge of having a few vocal people can also come up during Retrospectives. You might have a few people who have long-running problems that they're always concerned about or that there isn't a right solution for. They must have the chance to bring these things up, but if there isn't a correct answer or they are the only ones who see this as a huge issue, then it's not something that the team should be taking on.

Key Take-aways

- It's essential that everyone can voice their concerns and feel heard.

- Not all problems should be acted upon.

- Often it is essential to check in with other people on the team to see if they see the same issues or not.

- No matter what the feedback is, listen carefully, don't cut them off, and thank them for the feedback.

Enlist Some Allies

Regardless of how your company has been set up, there are managers that work alongside your team (often called Development Managers). As the team moves forward, these people can help or get in the way, so you don't want to leave it to chance. Managers or executives that work with the team will be valuable allies if you let them, and they usually have opinions as well as a background in the current and historical strengths and weaknesses of the team and how they performed in the past.

Of course, you may not necessarily agree with their assessments, or their insights may have been valid at one point but may no longer apply. Still, it's vital that you ask and listen carefully to their opinion about how things have worked and the strengths and weaknesses that they see. By listening, you're showing that you value their opinion and genuinely want to work together to help the team improve. Ultimately, their opinion matters, so you'll want to take notes and capture it. Of course, taking interest and notes won't be enough if you don't act, but we'll get to that later.

Along with a general assessment, every manager also has areas that they key into and will always focus on. Often these areas are based on massive failures or successes that they have had in their past, and professionally they know at a core level that these areas must be working well for anything to be successful. I once worked with a manager who used to work in customer-facing applications where having detailed documentation was critical.

Even though the teams I was working on were strictly working on back-end systems consumed internally, this manager was concerned and focused energy on creating the very detailed documentation that end users would expect. These concerns may or may not be challenged in your current situation. For your success, it's critical to understand what these areas are because even if everything else is going right, these managers won't truly see the team as being successful until these hot-button issues are working well. Fortunately, because they are passionate and have deep convictions about why these areas are essential, it's easy to get them to open up about them. Make sure to reserve a little time, because once you sit down and directly ask them "What areas do you see as key to the team's success and why," you're going to make an ally, but you'll need to be patient and let them talk. Of course, you will have your own assessments of the current challenges and areas that are key to the team's success. Because you are creating a partnership, you'll want to share those with the manager so they can understand that you have thought deeply about and sized up the situation.

Having an ally is very useful, of course, because you have someone else to help you change behavior and make improvements. An ally is also a second pair of eyes and ears, and a second mind to bounce ideas off or help you think through ideas when you have them.

Getting an ally won't do you much good if you don't keep them, so you'll want to follow through on that partnership. If things are going well, it may be something that you only need to do occasionally, but if the team is struggling, you'll want to be collaborating to make progress anyways. Here are the ways that you can keep a strong working relationship:

Share Information: When you hear about program, project, or other massive, impactful changes, be sure to share what you've heard with your allies. They may not have listened to the information that you have, or they may have additional details that you don't, so keeping each other in the loop helps you prepare for changes that will affect the team.

Share Insights: As you discover new things and weak spots for the team, allies are great people to bounce those against because they often have additional information or approaches to getting things fixed that you may not have thought of.

Support/Acknowledge Their Hot-button Issues: Mention or give visibility into ways that you and the team are improving or keeping a focus on the hot-button issues that your allies care about. If things are going well, this can be done with a quick mention or just a reference on how a fix or improvement will also help in the hot-button area.

Ask For Help: Being wanted is essential for everyone. Asking for help is a significant acknowledgment for your ally because you are demonstrating that you believe they can provide support and advice, and this helps them feel good about what they are doing.

Key Take-aways

- Having allies will be crucial to your success in your role as a Scrum Master.

- Providing value and keeping your commitments is critical in being able to keep your allies.

- Providing information, insights, and support for your allies and their causes will get them to do the same for your causes.

- Make sure to use your allies and ask for help when you need it.

Being in the Organizational Loop

To serve your team, you and your Product Owner need to be looped in and getting the latest information about where the product and company are headed next. Of course, some of the data is preliminary. Ultimately the company/product/feature may end up taking a different direction, but having a heads-up is critical because it will allow you and the Product Owner to start thinking about these future changes, thinking through how the changes fit or do not fit into current efforts/systems/architecture, or proactively heading off decisions or directions that might not make sense in the future. These future directions can also help you make smarter choices today, which could dramatically save you time down the road.

Recently I heard that the automation tool that had been the standard for years in our organization had grown too many drawbacks, and a replacement was being considered. Our team had come to the same conclusion a quarter earlier but hadn't been able to get enough people to back an alternative at the time. Armed with new information, I give my senior test engineer a heads-up. This was important because our team had already concluded that an alternative was needed, and we had already made significant progress towards a better solution. Our engineers were able to find and engage with the newly appointed group tasked with a replacement, show them the benefits, and progress we had already made, and get our solution adopted across the program (instead of seeing the future go in a different direction).

Key Take-aways

- Being connected to the organizational grapevine gives you and your Product Owner advanced warning about potential future changes.

- Understanding potential changes allow you and the Product Owner to prepare and influence their direction, saving your team time and effort.

Make an Ally of Your PO (Sidestepping Conflict)

I had a Senior Scrum Master role at a prominent tech company. I went through introductions, and within the first several days, I took my Product Owner to lunch to learn about each other's experience, strengths, and weaknesses. Over sushi and beer, I talked about wanting to partner, and how I thought we could complement each other's skills to represent the team well. Because he was new to the Product Owner role and wasn't given any formal training (very common, unfortunately), I volunteered, and he agreed that I would help and provide a little guidance with some of the Product Owner tasks that he wasn't very familiar with. After several months on the job, he went out of town for several days, and I filled in for him at several ends of Sprint ceremonies. When he came back, he heard about how smoothly things had run while he was gone, and he started to get a little worried that I was trying to replace him. I wasn't interested in his job, and while he was gone, I found some time to modify a Demo template that I used in a previous job with the current company branding. So, fortunately, I didn't have to tell him I didn't want his job. I showed him by setting him up for success. The next day he walked out in front of 32 teams, introduced our team, and launched into his presentation. Using the slides I provided, he explained the business value produced in the Sprint, the strategy around the priorities, what our future focus would be, and how it all fit into the product roadmap. No one else even came close to what he delivered. He sat down to glowing praise, and I intentionally sat on my hands and said nothing the entire presentation. That was the end of any tension between us and any fear that I was angling for his job.

The Product Owner is one of the three pillars that make up a functioning Agile team, so the role is critical for your success. No matter how good the Scrum Master is, if the work isn't being defined and broken down into good Stories, then the critical work isn't being prioritized, the business doesn't feel like they have an inside view into the team/progress, the customer isn't being represented, and your team is going to struggle. All good teams have a partnership between the PO and the Scrum Master. There will be plenty of times when the Scrum Master won't be able to make a Standup, or a Product Owner must skip a User Story Grooming meeting to meet with the business, so it's going to be essential to have a strong partnership with the PO.

The Product Owner role is also, unfortunately, a role that the Agile community hasn't traditionally trained or supported well, so there aren't as many Product Owners who know the role profoundly. Having a trusted advisor/partner relationship with your PO is going to be critical. Unless you are roulette wheel lucky, you will need to help your Product Owner grow into the role.

Product Owners (and especially new ones) can be intimidated or even threatened by the Scrum Master's role with the team, and that can quickly lead to miscommunication, distrust, and conflict. It's essential to establish the partnership and put their mind at ease about the role you're playing with the team. The significant points that you'll want to make (and I'll go over it in detail below) are: "I'm thrilled you're here," "Let's partner," "I can help you," "Let's back each other up," "Here is my goal and how I see my role," "Do you have any ideas on my role or how we can collaborate?". You'll want to customize these based on your situation, but here are detailed talking points that will serve as a good starting point. There is a certain amount of flattery because you'll want the PO to feel confident in their role and comfortable that you accept them and their abilities.

Introduction to the Topic: Thanks for coming. Since we're going to be working so closely together, I thought it would be useful to get together so you can understand my background and we can talk about how best we can support each other and partner while we're helping the team.

I'm Thrilled That You're Here: While I'm good at Agile, we need someone who has in-depth product knowledge and knows how to connect to the business to be successful. I just don't have that experience or expertise, so I'm delighted that the team has you for a Product Owner.

Let's Partner: I know that there will be plenty of times that I won't be able to make Standup, and I might have conflicts. I know that there will probably be times when you are stuck in meetings with the business and won't be able to attend User Story Grooming Sessions or other ceremonies. Hence, I thought that it would be useful if we partner so that if I can't make a Standup or you can't make a Grooming Session, we can cover for each other, and the team won't have a gap.

If there is ever a ceremony that you're going to miss, just let me know, and tell me how I can best represent your views, priorities, and the changes that you need.

My Goal and Role: As I see it, my primary goal is to make sure that you, both the PO and the team, are wildly successful. No matter what else happens, if I haven't done that, I have done my job. To ensure success, I'll provide structure and expertise in Agile and the Agile Ceremonies, support and back you up, and support the team. Does that sound like a plan?

Get Buy-in: Do you have any thoughts on how we can partner? Does this sound like a reliable approach?

Key Take-aways

- The Product Owner Role is critical to the success of the Agile Team.

- To make things work you want to make an ally of and work closely with your Product Owner.

- As Scrum Master and Product Owner, you need to back each other up and fill in for each other when one of you isn't available.

- You want to ensure that the Product Owner doesn't see you as someone trying to compete with them, but rather someone helping them.

PART 3 – PRODUCING
QUALITY PRODUCTS

Shielding the Team

A little understood but critical role for the Scrum Master and Product Owner is the need to shield the development team. When they are allowed to focus, good development teams can create fantastic value. Unfortunately, there can only be one thing at the top of the priority list, and everyone else would love to find a way to get their priorities done even if it means using alternate and unofficial routes. Getting a new feature rolled out the door to stay ahead of the company's competition might be the highest priority. Still, a customer support manager who has a massive bonus on the line and a customer who has been complaining about a bug in a seldom-used feature might see things very differently. Everyone has their own priorities, and they all want to influence your team in their direction.

In large organizations, there tends to be a lot of politics and sausage-making around how the strategic direction for the business is decided and who gets to decide it. Deciding how and why million-dollar budgets will be spent and how the company should pivot their applications and customer tools requires a lot of debate and negotiations. The challenge for development teams is that all of this is very distracting and ultimately demotivating.

In the early days of my career as a Scrum Master at a large bank, I thought that I should keep the team informed of everything that I knew or found out about the future direction of the product we were working on. One week during Sprint Planning, a senior dev looked at me and said, "Why should I work on this? The business has canceled and restarted this project four times already. If I start working on this, they are probably just going to cancel it again in two weeks, and I'll have wasted all my time." Right there, I realized my mistake. Instead of giving him and the team every little bit of product gossip I had been collecting, I should have only brought them information about work that had been committed and detailed.

Key Take-aways

- The Product Owner and Scrum Master need to shield the development team from distractions.

- Not everyone in your company may prioritize the work the same as the Product Owner will.

- It's crucial to shield your team from the drama and sausage-making that goes on in the business so the development team can focus on coding.

No Untracked Work

Starting at a new data visualization company and supporting an existing team, I did my homework by looking over the backlog and the User Stories to look for the quality and completeness of the writeup, as well as to get a background in the work they were focused on. During the next Sprint, I noticed a difference between the Story points in the User Stories and how many Story points were allocated for the team's velocity. I was told that the team allocated 10% of their time (roughly eight Story points) for product support issues, so there wasn't a need to track that work directly; it was just built in.

I checked in with a senior dev to see what type of production support issues they typically encountered, and I listened to a long list of complex and ongoing support that sounded like it was a significant time commitment and not something that was a minor side project. I decided to test my theory out, and I talked with the two developers who did most of the production support work. I asked if they could keep a short one-line log capturing the production support tasks that they did during the next Sprint and how many Story points of effort each item was worth once it was completed. At the end of the Sprint, I came by and collected the lists, got the last-minute updates, and went off to add up our actual spend for production support.

The team was shocked when I shared that they were spending 37% of the team's time fixing production support issues and that they were a Story point total three times greater than the

10% uplift that they had been using for years. That information clicked with the team, and a lot of other challenges that they were facing started to make sense. They were struggling and were consistently not finishing up the committed Sprint Story points. I'm sure that at one point the 10% uplift was a number that made sense at one point in their past, but over time, as there was more and more functionality released, there was more and more that needed to be supported, and the need had slowly grown to become a monster impact. Problems can spring up quickly, and just keeping a general uplift won't capture problems when new and impactful issues spring up.

The best approach is just to teach the team to capture work items that take more than an hour. You can change the granularity of how you measure it, but anything that is taking any significant amount of time should be captured. It works best to make the tracking easy, since you don't want to be asking developers to spend 15 minutes to create a full formal User Story to track an hour's work of production troubleshooting, so we like to use what we call Tracking Stories. These are lightweight stories that are just used to track work that is short and easy to understand, and the only thing that's required is a title, a tag of "Tracking Story," and a time estimate for what was spent. For instance, the team that I currently work with tracks bugs in a tool called Bugzilla (an open-source bug tracking tool). With all the details and notes captured in Bugzilla, the Tracking Stories that we create don't need to be detailed. These bug Tracking Stories usually only contain the Bugzilla bug ID number, a link to the issue, the title of the bug, and the number of Story points. With this information, you or the Product Owner can easily collect these Stories at the end of the Sprint and detail how many Story points were spent. It works best to have the team capture the effort after they have done the work because with production support work, they aren't going to know how long it will take, and the critical part is to be able to capture the effort spent accurately.

Key Take-aways

- You can't know how to manage your work until you can accurately understand where the work is being done.

- When tracking work, the User Story (Tracking Story) doesn't need all the details that it would have for feature work.

- It is easy for maintenance work to grow over time to become more significant than you expect it to be.

Saying No Without Saying No

In large companies and corporations (especially those with a more formal business process), it is prevalent for them to have an informal economy. By informal economy, I'm not talking about anyone selling hot Honda Parts to beef up their street racers, but a way to get around the normal work request processes that are there to vet and approve new items. Whenever someone gets around the official work request process and gets something done in half the expected time, the more this encourages them to try again next time, and the more likely they will get a "yes" from someone on your team who has already helped them out in the past. After repeated success, the new workaround becomes the default way of doing work, and the checks put in place to ensure that your team is working on the right things get bypassed. This leaves the team with less time to get critical work done because they are busy doing work for someone else.

Yes, it's entirely possible that some or maybe even all of the requests coming in are important, and that the team would end up doing the work anyway, but that still creates a problem because you can't understand this underground economy, the costs that you and the team are paying for it, or whether it's priority work that's more important than what you are working on currently.

If you are careful with how you ask and do it in a very casual, non-judgmental way, it's effortless to uncover if your team is being affected by the underground economy. All you need to do is to casually add the following at the very end of your next Standup "Hey, I'm just curious... Is anyone helping or doing any little tasks to support other teams?" You may need to ask once a week across several weeks, but if the team doesn't think you'll be angry or grumpy, you should be able to uncover what type of requests are being made of your team. That, of course, doesn't mean that you'll be able to understand how much of an impact this is having, because getting a handle on the amount of work will be much more challenging than finding it in the first place.

The next step is to talk with the people who are fielding these requests to understand how often they are coming in, what type of work they are, and who is submitting the work using this back door approach. It's possible that the requests are small, don't require much time, and are being helpful to the organization, but it's also possible that they are, in fact, a hidden problem.

It's hard for someone who has been helping someone else to just cut them off cold turkey by saying "No, no way, I'm not going to do that!" because they did it last week. The most effective way is to explain to your team member that if the request makes sense and should get done, that you'll approve it, but you want to have visibility into the requests, so they should be routed through the Product Owner, or yourself if the PO is away. Then you teach them that they don't have to say "no," they can just say the following:

"The request makes sense; I can see why you need that, just route it through my Product Owner so he knows where I'm spending my time, and I'll get it done for you."

This approach is significant because your teammate doesn't have to say "no" and they are leaving the person being hopeful, but they aren't saying "yes," and they know that if the Product Owner needs to say "no," then they don't need to be the bad guy. In the end, the phrase is more effective, more likely to get used, but still routes the request back through someone who can adequately judge the priority of the request against the current work.

Key Take-aways

- Side work can be a major distraction for a team and ruin productivity.

- Find out how much side work is an issue by asking your team (in a kind and non-judgmental fashion).

- It's easier for individual team members to redirect requests to the PO instead of having to say "no."

Setting (and Keeping) Product Standards

Setting expectations is helpful because everyone knows what standards your team keeps. If you're new to the team, or you haven't reviewed it recently, going back over your standards documents is a useful exercise and something that you should do every six months. Without standards in place, everyone in and outside the team can have different ideas about what "done" should look like, and that leads to confusion and conflict when one person considers it done but another doesn't.

The Definition of Done: Is one of the primary must-have documents for your team. The Definition of Done is a generic set of Acceptance Criteria that should be applied to every User Story. Of course, there will be times when something on the Definition of Done doesn't apply to a Story (like a research Story that doesn't produce code won't be checked into source control), but if the rule is generally useful, go ahead and include it. For instance, the code for any Story that has been completed should have a code review. It may seem obvious once you think about it, but some things aren't apparent, so it's best just to spell them out and capture them. When the team is grooming Stories, the Definition of Done will allow your team to concentrate on the Acceptance Criteria and not have to reinvent the wheel every time they write a Story. To complete a Story, the team will meet the Definition of Done, which provides the foundation, along with the additional items spelled out in the Acceptance Criteria that are unique to that given Story. It works best to have the Definition of Done copied into your User Story template so that, when a new Story is completed, there is a single location that lists out everything that needs to be accomplished for a given Story to be finished.

The Definition of Ready: Is also an important document that spells out what needs to be in place before the team is ready to consider working on a User Story.

Early in my career, I was surprised by a short notice meeting with an important stakeholder from sales about some hot new features he needed. My PO was on vacation, and I was filling both the PO and Scrum Master roles. After the first meeting, the stakeholder went silent, never answered questions, and didn't deliver other information that was needed. I made the mistake of writing up the feature and getting it added to the backlog for the upcoming Sprint. The same stakeholder who wouldn't provide the needed information did show up at the Sprint Demo to loudly deliver scathing criticism of what was delivered. If you are getting Stories directly from marketing or the business, there often won't be the enough details. The team could, of course, make assumptions, but I guarantee that if you do, you'll have unhappy customers when the team guesses wrong. The best way to be successful is to have a firm and well-published Definition of Ready and get your PO to practice the following phrase: "I can see that this is important to you. This Definition of Ready details the information that we need to be able to do this work. Just give us these details, and we'd love to try to get it into this next Sprint." You must train your customers to provide what you need to be successful, or they will give you sparse information and be unhappy when you haven't delivered exactly what they wanted. If your customer can't spend 30 minutes to give you the details, then they either didn't need it that badly, or they hadn't thought through what they wanted. Either way, you're not going to be successful until you get enough information.

Key Take-aways

- Standards are an essential tool in establishing the expectations and quality of the product.

- The quality of the work and your team's reputation will suffer if people have different expectations about what is being delivered.

- The Definition of Done is a list of generic Acceptance Criteria.

- Most, but sometimes not all, of the individual items on the Definition of Done will apply to any given User Story.

- The Definition of Ready keeps teams from wasting time on Stories that aren't ready to be worked.

- Having a Definition of Ready shields, the team from customers who want things but won't spend enough time to tell you what they want.

- Trying to guess what your customer wants is a losing battle and one you want to avoid.

The Motivation of the Midnight Phone Call

There are plenty of organizations that migrate slowly over to Agile, with many never making it very far. Often these halfway implementations will keep the most convenient and often destructive habits, and one of the most tempting is to have sustaining or production support teams. The logic behind having these teams is that your best and brightest forge ahead building extraordinary functionality without being distracted by the tedium of dealing with bugs from customers or midnight phone calls when the website or product is having problems. While this is tempting, the reality of what happens creates significant problems for the business.

The Challenge of Having Large Production Support Development Teams: The primary product development teams are always under pressure to produce enhancements and new products, leaving someone else (anyone else) to deal with grumpy customers and late-night calls when things break. When someone else must deal with the consequences, the quality and stability suffer because there just isn't the motivation to make it right.

The other problem is that the best and brightest move over to the product development teams and the production support team, or sustaining team, usually ends up being a C-level team (of course there are some talented people, but if you average it out, they don't have the same horsepower). Not only does the team have less talent, but they weren't involved in building the product in the first place, so they aren't as familiar with how it works, nor the trade-offs or weak points.

The product development team is, of course, very busy working on the next release, and then the next project, so they don't have the time to document things very well and are too busy for a proper turnover.

All the communication gaps and the lack of a quality mindset means that when you have separate production support teams, they are generally lacking in information and working with a lower quality product. Ultimately, it's your customers that suffer, which drags down the reputation and stability of your product and in turn erodes your profitability.

A Better Approach: When teams and management know that they'll be the ones who will get the phone call if something critical breaks in production, then calculations become different. I'm not saying that making this transition will be easy – if the quality is low currently, it's going to be a little painful, but just imagine what your customers are going through. If you are bringing responsibility back into the team, I would recommend having a good turnover with the sustaining team and spending a couple of weeks to go through and harden/fix any of the broader issues that are impacting customers currently. You'll need to spend a Sprint to do this, but having a quality product is essential, and this will allow your development team to own the stability of the product going forward.

Key Take-aways

- When someone else is going to deal with calls in the middle of the night, there isn't as much motivation to build in quality.

- Having a separate group to support production leads to an information/understanding gap because they weren't the ones that made the product.

- For the best quality, bring production support into the team that makes the product. It does take more time and effort to build in quality, but your customers will be much happier in the long run.

Keeping Quality in the Team

As organizations take steps towards Agile, it's not uncommon for them to have a separate and distinct QA team that acts as a single point of contact for all the quality work within an organization. While this may be convenient for people management, let's look at how that works out when it comes to quality, since that is what those people are in the organization to do.

When the work is being done in one group and then turned over to another (the QA group), that gap creates a significant barrier to communication. Your QA engineers won't have as good of a view about how products or features are being put together in the first place, and what information they do get will be received later in the product cycle, so they will have much less time to use the information. By being on a different team, QA also doesn't get included in all the early design discussions, trade-off decisions, and the context that leads up to the product, making them much less prepared to do an excellent job of testing it.

Along with a gap in communication, people on different teams may feel that their motivations aren't aligned. If I'm a developer who takes pride in my code, then I probably won't take the time to help someone on another team that's going to tear in and try to break my brand-new feature. If they are on my team, I'd probably help them. But if they are on another team, I might be tempted just to provide minimal communication and focus on my next task.

To be successful, you'll need to bring QA into the team. If that's not possible, then you'll want to do everything that you can to pull those external people into your team process to try to eliminate the communication and motivational gaps. This might include:

- Getting the QA and Dev teams to sit next to each other.

- Setting up design reviews so that QA engineers can understand how to test new functionality.

- Holding a joint Standup.

Key Take-aways

- When QA is a separate group, they don't get the understanding and background with the product like they would when involved at the beginning.

- Even with the best of intentions, communication between groups hampers coordination.

- When in separate groups, the motivations of individual team members or the entire group may not be aligned, causing the product quality to suffer.

Making the Right Thing the Easy Thing

It doesn't matter what type of work you're doing – ultimately, doing a quality job takes more time and effort. Just as products get more complex over time, the amount of work that it takes to produce a quality version of the product will also increase. If you don't invest in automating some of the team's work (usually testing), the tasks for you and everyone on the team will increase in number and complexity and your team will have less face time doing the actual coding that they need to roll out a new product.

You can see this in action at large, long-lived companies. Applications that were the latest and greatest, most stable code available get tweaked, added onto, and updated until the original code becomes unrecognizable, unstable, error-prone, and difficult to maintain or update. At some point, it becomes evident that so much money is being poured into the broken and limping application at the mechanic's shop that it would be smarter to make a large spend up front to build a new version.

So… Doing a quality job up front is critical, and to give your team the maximum amount of time with the code, you'll want to automate and simplify everything else that you can. The best way is to make a list of all the work and items that are being created and see how you can automate them. On my Kanban boards, I like to have a column for code reviews, which makes it an automatic next step and not something that the team has to remember. Of course, the whole team knows that code reviews need to be done for everything that we produce. Everyone knows that it's expected that they all pitch in to look over teammates' code, but having a column on the board helps because it's visible. If someone forgets to do a review, then at the next Standup, everyone will see that same Story stuck waiting to be looked over. With this system, I don't need to nag anyone about doing code reviews. It's transparent, and that helps make it easy and more likely to work well by default.

Here are some of the significant areas that you might consider building in:

1. Build in a code review column into your Kanban board as a visual reminder for doing the code reviews.

2. Create a comprehensive template for User Story creation. It's easier to ignore or delete sections than it is to create them, so just have the details and use what you need. The best is when you can set it up so that any new User Story that gets created spawns with the template already in place.

3. Set up a reminder to send out User Stories before Story Grooming Sessions.

4. Set up a reminder to close out your Sprints.

5. Create a spreadsheet or template to calculate your velocity automatically and remove Story points for people taking days off.

Key Take-aways

- Over time the amount of time your team needs for day-to-day tasks will slowly increase, leaving them with less to get stuff done.

- Building the process so it's easy to do the right thing will save time and help your team to work smarter.

- Investing in automation is sleek and allows you and the team to focus on value creation.

PART 4– PUTTING IT ALL TOGETHER

Being an Agent for Improvement (Instead of Change)

Most people aren't excited about changes to their world or how they work, and some people can be downright hostile. People who are hostile or suspicious of change will make adopting even the best-intentioned changes more difficult by being critical, nit-picking details, voicing dissent, gathering support to keep things the way that are, sabotaging the change effort, or even refusing to participate.

The good news is that while change can be uncomfortable, most people like improvements that make their life or work easier. The secret of making progress is to stop talking about changes and start finding improvements to genuinely make life easier for people on your team.

As a Scrum Master who is new to a team (or just trying to gain some positive momentum), the first handful of improvements will be the most important. Successful improvements that save time, effort, or resources build trust in the Scrum Master's ability to truly help and create appetite for more improvements. Success also lowers the team's skepticism of future improvements and helps plant the idea that there may be a better way for the team to be working than the one it's using now. Failure for changes to help the team will have the opposite effect, so it is critical to establish success in the beginning.

Pace: Getting early success to gain the team's trust and establish the Scrum Master as an agent of improvement is critical. Don't rush – take the time to really understand what is and isn't working well for the team. To do this, the Scrum Master should interview each person on the team individually (casually if possible), then leadership, and then stakeholders.

Through these interviews the Scrum Master will be able to identify the issues the team is having, how people outside the team view its performance and issues, and the differences between how individual team members view the team's challenges.

Having this wholistic background will make it much easier for the Scrum Master to support the team to make forward progress. This understanding is especially important on teams where many team members may be holding their thoughts and opinions back.

Key Take-aways

- Change is uncomfortable and will often be meet with resistance.

- Improvements aren't met with the same hostility if they make life easier for those who will be adopting them.

- A Scrum Master who establishes themself as an agent of improvement will be able to get more done quicker and with less resistance.

- Getting early success for improvements is critical to establish the Scrum Master as an agent of improvement, so take your time and make your opportunities count.

- When working with an unfamiliar team, it's best for the Scrum Master to conduct casual interviews with the team and individual members to understand what challenges the team might be encountering.

Order of Change

Now that you have a better understanding about what needs to be in place in a mature Agile program, the next challenge is to implement all the things that we've learned in this book. In a perfect world, this would be easy, but human nature makes it a tougher assignment.

The challenge is that revolutions are messy. When you're trying to make such a large and dramatic change in the way that your team works, you're going to run into resistance. By nature, people aren't comfortable with change. The larger the change is, the more it is going to be resisted.

When trying to create positive change in a team or organization, there is a natural order in which to take steps. The larger the change, and the more out of order the change, the more hostile the reception will be.

To be a successful Scrum Master you will need to be strategic, patient, and careful about putting changes in place. While it is possible to do things differently, like most things in life there is a certain order in which things should be done. It is possible to have some overlap between these different phases, but trying to jump ahead too far will make success extremely difficult (if not impossible). Here are the major stages in the order which they work best.

Gain an Understanding: In an ideal world, an Agile team would be able to identify all its own problems, come up with its own solutions, and move forward with them. Of course, we don't live in an ideal world, which is why the Scrum Master exists in the first place.

The role of a good Scrum Master is to help support the team to move towards success and move towards the ideal. The best way for the Scrum Master to do this is to examine and fully understand the situation and challenges that the team is facing while also understanding the current team dynamics.

This is accomplished by talking with team members, managers, and stakeholders to understand how each person views the team's health, strengths, and challenges.

With a deeper understanding of the situation, the Scrum Master can then make intelligent choices to support the team moving forward, regardless of where on the Agile or team maturity cycle, they are.

Got Team?: A functioning development team is an amazing and dynamic entity capable of amazing things. As a Scrum Master, before you make plans about anything else, you need to answer the biggest foundational question: "Do you have a team? Or do you have a group of people with a team name attached?". A team is a group of people collaborating to deliver on a common goal. If the team isn't working toward a common goal, or not working together, then they aren't yet a team, but rather a group of people who are still forming. The answer often isn't a binary yes or no, but it's critical to have an answer to this question before next steps can be decided. If it's a team, then there is a foundation to build upon, and if it isn't a team, then that needs to become your top priority.

To figure out if it's a team you need to answer the following questions:

- Is everyone working towards the same outcomes (or at least in the same direction)?

- Is it safe enough for people to put forward new ideas or disagree? Is it psychologically safe?

- Are ideas and input being put forward by most of the people on the team, or just one or two individuals?

- Are people comfortable being themselves?

- Is the body language of team members open and relaxed when everyone is together, or are people tense and closed (arms crossed, leaning back)?

- How does the energy change when the group gets together?

- Are team members willing to work towards the team's goals instead of their own?

Until there truly is a team, conversations with the group will be dominated by a few individuals, making it difficult to really understand what the team needs. If too many of the questions above don't have positive answers, then trust and team building should be the primary focus.

Key Take-aways

- The Scrum Master's role is to help the team succeed.

- To be able to help the team, the Scrum Master must first gain a deeper understanding of the team's strengths, challenges, and dynamics.

- Interviews with the people on or interacting with the team are the best way to gain a better understanding of the team.

- If the team isn't collaborating, or working toward common goals, then it is a group of people and not a team.

- Until there is a team, conversations and Retrospectives will be dominated by a vocal minority and true progress will be difficult.

It's Not Just What's for Dinner

Anyone involved with Agile knows about the Retrospective ceremony where the team gets together to identify areas to tweak and improve. Mature Agile teams make tweaks and improvements on their own and use the formal Retrospective ceremonies as safety nets to catch anything that might get away. Mature teams will also continue to tweak or find replacements for solutions that don't work as intended.

To establish team improvement behavior, and to establish the Scrum Master as an agent of improvement, it's best to find small, isolated, low risk, big reward changes that can be made (or just a low risk change if a juicier target isn't available). For instance, one developer started several weeks after I started at a new company. There wasn't a good onboarding document describing the tools and environment that would need to be set up for the developer to do their job, so an obvious solution was to have the new developer to gather the information and create an onboarding document for future team members. When talking with the team, the Scrum Master should bring up the challenge, how or who helped identify the issue, and then ask the team how they can solve the problem. If the team puts forward a potential solution that has the backing of a majority of the team, then acknowledge the idea and the person who proposed the fix and set a reasonable timeframe for the proposed solution to be put in place and tried out. It is important to point out that the solution will be tweaked or removed if it doesn't work. The conversation might look something like this:

Scrum Master: When I was talking with Jason, he mentioned that one problem that we continually have is when [issue] happens. I can see how this would create friction for us. How can make an adjustment so this isn't an issue?

Engineer: What if we put [potential solution] in place?

Scrum Master: Chris is worried about [complication] but everyone else seems to think that this solution could work. We have another week before the Sprint ends. Let's put this in place, and we can try the solution out over the next Sprint. If the solution works, we can keep it in place. If it doesn't, we can revisit it and find a different way forward.

Pace: Even if they make life easier for everyone on the team, there still is a limited appetite for change. If the Scrum Master and team try to modify too many things, people will quickly become uncomfortable and start pushing back. Start very slow, focus on items that everyone on the team sees value in fixing, and work to make solutions stick before moving on to new items. After establishing trust, clear benefits, and a reasonable pace, the team will then be in a place to adopt the formal Sprint Retrospective process without major resistance. A typical progression might look something like this:

1. Week 3: Minor, super easy fix. (Maybe improve the query used to generate the board during Standup.)

2. Week 4: Minor, easy fix. (Maybe provide a nice-looking template for Sprint Demonstrations.)

3. Week 6: Small, big benefit. (Maybe remove an unproductive meeting.)

4. Week 8: First Retrospective.

5. Week 9: Minor.

6. Week 10: Second Retrospective.

Key Take-aways

- The best teams make changes and improvements regularly, not just during Retrospectives.

- Low risk, big reward changes are the best way to jump-start a culture of improvement and to establish the Scrum Master as an agent of improvement.

- Finding a solution to a problem that has already been identified by the team is a great way to get early wins.

- Making too many improvements too quickly will also make many people uncomfortable, so space them out.

- Successful improvements will empower the team to suggest and implement more of them.

When You Need to Slow Down

Young inexperienced teams usually have a lot to learn. They have large challenges, and often haven't developed enough confidence and trust in each other to tackle big problems by themselves. In the early stages, a good Scrum Master is firmly engaged with the team, acting as a leader to help the team evolve toward a prosperous and productive future.

This is done by using a servant-leader approach, but the balance between the two sides will change. As a team becomes tighter, more effective, more Agile, and able to function as a single unit, a good Scrum Master should adjust their engagement to match the team's level of effectiveness.

A mature team firing on all cylinders and being effective will still need occasional guidance, support, teaching, and everything that a good Scrum Master will provide, but they probably won't need as much direct engaged leadership as an immature team would need.

The Scrum Master should be constantly assessing the situation and the maturity of each team that they work with, adjusting regularly. If an area is going well, then acknowledge the good work, the success, or the leadership that has taken place, and step back in that area to let the team take a stronger lead. Your goal as a Scrum Master is to facilitate growth and independence so you can step back into more of a supporting role once your team is really working well together.

Key Take-aways

- As teams mature, the Scrum Master's involvement should ease off to allow the team greater self-management.

- Teams will always need the guidance and support of a good Scrum Master.

- Teams will have different levels of maturity and self-management in different areas.

- Assess and adjust as needed because the maturity and success of self-management will change regularly.

CONCLUSION

Congratulations! By reading this book you've taken a bold step forward to improving your Scrum Master skills, your career, and the success of your company. Let's review the content that we've gone over:

- We talked about every Agile ceremony that the Scrum Master will be attending, how to deliver, what to avoid including, and how to get the most value out of a good Sprint Retrospective.

- We detailed the value of well-written User Stories, how to refresh out-of-date Stories, and then gave you the information and templates to help you deliver them.

- We detailed how to get your team the visibility and recognition it deserves by delivering top-notch Sprint Demonstrations and providing Public Relations for your team.

- We talked about how and why to avoid short-term thinking, having QA outside of the team, and poor quality and untracked work.

- We talked about the importance of saying "no," and why it's better to say "No, we don't have room for that in the next Sprint" than to say "No, we didn't get that done" or "No, it's not stable."

- We talked about how to build collaborative and successful relationships with the key people that partner with the Scrum Master and work on the development team.

- We talked about how the Definition of Done and the Definition of Ready are key documents that are needed in the day-to-day operation of an Agile development team.

- We talked about the importance of keeping quality in the team and the benefits of having the teams that develop a product also be the ones that fix production bugs.

Being an effective Scrum Master is an art as well as a discipline. It's a role that takes balance, patience, a thick skin, and the willingness to ask questions and to dig into why processes, organizations, and people work the way that they do. I love the profession because it allows me to have such a large impact on the quality, engagement, happiness, and success of my teams' lives and careers.

You are a teacher who helps spread new ideas, a guide to all things Agile, a coach who encourages effort and cheers for wins, an advocate who sticks up for the team and what they need to be successful, a Public Relations expert to help spotlight the amazing accomplishments of your team, a leader who helps push the team to keep their commitments, and a counselor to help diffuse conflicts. Putting too much energy and focus in one area will see diminished returns. Doing the job well means maintaining a balance, being patient, and keeping the health of the team in constant focus. To be successful as a Scrum Master, you need to care about the people and team that you work with, not just the efficiency, the customer, and the product.

Thank you for coming along on this journey. I worked hard to capture the wisdom and experience gathered during my career to help you become successful as a Senior Scrum Master or Agile Coach. I sincerely hope that this book has been useful for you, and I'm glad that you've been able to come along with me on this journey. The next step for you is simple. Whether you want help or want to go it alone, it's time to roll up your sleeves and get started.

Don't over-think this, and don't try to improve too many things at one time. Make decisions about what you should focus on, make improvements, and keep going. The transformation won't happen overnight. It will take time, but you've learned everything that you need to be a Rock Star Scrum Master. You just need to get started.

How To Get More Help

If you do decide you want our help transforming your career or your organization, just reach out and let's sit down and chat.

We're good at what we do, just like you're good at what you do. We have had the time to become experts in the Agile space and if you have something to offer the world or your organization, then we want to be the ones to amplify it for you. You can find us at AdvancedAgileExecution.com/call.

About the Team

Advanced Agile Execution is a collaboration between Matthew Kramer and Tanja Diamond. Our mission is to help you go from being an average Agile user to an Advanced Agile BOSS!

Matt started his tech career in Seattle, WA in 1998 working his way up from QA Engineer, QA Lead, SDET, QA Management, Program Manager, Scrum Master, and then an Agile Coach. Matt has worked in small startups, Fin-Tech, and Fortune 100 tech companies.

In 2018, Matt completed the construction of a two-thousand-square-foot barn which started with the initial layout of the foundation, framing, and plumbing and finished with the wiring (he did get a little help with the wiring).

When he isn't building barns, furniture, or Agile Programs, Matt volunteers to help young adults land their first tech jobs and grows tree starts to distribute them in the community.

Tanja Diamond is a Maverik leader and has been a Business and Life Strategist for over 38 years. When she isn't out leading innovation in her clients' lives, she is saving lost and injured animals.

Be Amazing,
Matt Kramer & Tanja Diamond